TRUSTING ROBIN:

The cover picture shows an English robin trusting at the hand of man. In the chapter on prayer, I have noted an experience where a lamb trusted, and was probably saved from becoming a fox's meal. (Some have said that I may have robbed the fox of a meal – but let us not look at it that way!) This book is written in the hope that the reader will come to trust God in a deeper way than ever before. (Let not the devil make a meal of us – *do* let us look at it *that* way!)

THE CHRISTIAN CHALLENGE

From ALPHA to OMEGA

By
Brian Henson

AuthorHouse™ UK Ltd.
500 Avebury Boulevard
Central Milton Keynes, MK9 2BE
www.authorhouse.co.uk
Phone: 08001974150

© 2007 Brian Henson. All rights reserved.

No part of this book may be reproduced, stored in a retrieval system, or transmitted by any means without the written permission of the author.

First published by AuthorHouse 9/25/2007

ISBN: 978-1-4343-2529-7 (sc)

Printed in the United States of America
Bloomington, Indiana

This book is printed on acid-free paper.

To the Glory
Of God

ACKNOWLEDGEMENTS

First and foremost I acknowledge that without God's grace I would not have been able to write this book. He has brought me through many scrapes with disaster, some of which are recorded within. He also allowed me many experiences, which prompted these thoughts, and led me to put them on paper. I am grateful for all He has taught me - and is still teaching me.

Secondly, I am grateful for the encouragement I have received from my brothers and sisters in Christ. I am particularly grateful to the Reverends Clinton Morgan and Joe Kitts who encouraged me and confirmed that this book should be printed.

CONTENTS

PART ONE From Salvation to Discipleship

Introduction   xiii

Chapter One
What Really is a Christian?   1

Chapter Two
Being Filled With The Holy Spirit   7

Chapter Three
Claiming the Promises of God   13

Chapter Four
Doing What God Asks   15

Chapter Five
Comforting Others   23

Chapter Six
Prayer   29

Chapter Seven
God's Protection   37

Chapter eight
Being a Disciple   51

Conclusion   61

PART TWO From Deception to Deliverance

Introduction (Part Two) 65

Chapter nine
How It All Started 67

Chapter ten
Salvation 71

Chapter eleven
Some Thoughts 77

Chapter twelve
More About Thoughts 79

Chapter thirteen
The Spoken Word 85

Chapter Fourteen
The Written Word 89

Chapter Fifteen
What Hinders 91

Chapter Sixteen
Deliverance 95

PART THREE Two Prophesies and Epilogue

Prophesy For Today - No 1 105

Prophesy For Today - No 2 107

Epilogue 109

PART ONE

From
SALVATION
To
DISCIPLESHIP

All Scripture References are from
The New King James Version
THE HOLY BIBLE

INTRODUCTION

This book is intended for those who are already Christians, and desire to have a closer walk with God. I would like us to look at that walk as we continue as Christians. Just as Jesus is the 'Alpha'[1] and the 'Omega'[2], I feel that from salvation to becoming a disciple is like the Alpha and the Omega of our Christian journey- a journey that is a challenge indeed!

> *"I am the Alpha and the Omega,*
> *the beginning and the End,"*
> *Says the Lord,*
> *"Who is and who was and who is to come,*
> *The Almighty."*
>
> Revelation 1 verse 8

When we first became Christians we committed ourselves to a relationship with God through Jesus Christ. But are we now falling short of what God desires from us? If that is the

---

[1] 'Alpha' (A) - the first letter of the Greek alphabet.
[2] 'Omega' (Ω) - the last letter of the Greek alphabet.

case, surely this is not only to our own loss, but for non-Christians also? How many of us would be devastated to be told that our behaviour does not encourage (or may even discourage) others becoming Christians? Ultimately is hell their destiny because of this? Am I being too harsh? Are some offended by this question? Try to imagine facing Jesus and asking Him…. What would be His reply? How do you feel about my suggestion that He may not reply in words? How would you feel if I suggest that when you look into his face, wondering, you see the tears welling up in the corner of His eyes - then see a tear or two rolling down His cheeks? God loves us - **and** He loves those who are not yet saved. God wants none to perish! Are **we** helping or hindering their salvation?

At this point I should like to stop a moment to ask if you feel challenged by the above. If so, please do not give the credit to me! When I started this work I asked God to give me the thoughts about which **He** wanted **us** challenged. Indeed, the challenges are as much for me as any reader. Any praise belongs to God, not me - I am just His servant.

Chapter One

WHAT REALLY IS A CHRISTIAN?

I remember being asked in my mid teens, more than half a century ago, whether I was a Christian. I was still a schoolboy, and it was during the Easter holidays. It was, in fact, the Good Friday! I had been working on a farm to earn some pocket money and the man in charge of the dairy section asked if he could walk with me to the bus stop (he was the one going to catch the bus; I had a bicycle and was going to cycle back to the city). I agreed. That was **one** of the most important decisions of my life! During those few hundred yards to the bus stop Peter asked me that question. I remember responding, "I try to be." Of course, what I meant was that I tried to be good person. Then, that is what I understood was meant to be a Christian. Thank God that Peter did not leave it at that! He went on to say what Jesus had done for me! He explained that Jesus had died so that **I** could be forgiven all I had done wrong, that Jesus had died for **me**. I also remember that cycle ride which followed - I had been deeply touched and moved to tears. I then knew what I had to do to become a Christian. I accepted what Jesus had done, asked Him to forgive my sins and come into my life. That was **the** most important decision of my life!

It is said that when we ask The Lord into our life we have a new life – that we are reborn. I quote these words from St. John's gospel:

'Jesus answered and said to him,

"Most assuredly, I say to you, unless one is born again, he cannot see the kingdom of God."

Nicodemus said to Him, "How can a man be born when he is old? Can he enter a second time into his mother's womb and be born?"

Jesus answered, "Most assuredly, I say to you, unless one is born of water and the Spirit, he cannot enter the kingdom of God.

That which is born of the flesh is flesh, and that which is born of the Spirit is spirit.

Do not marvel that I said to you, You must be born again."'

Chapter 3 verses 3 to 7

But is asking the Lord to forgive our sins and to come into our life all that is necessary? Are we then good Christians? Admittedly we can claim to have responded to:

> *"Behold, I stand at the door and knock. If anyone hears My voice and opens the door, I will come in to him and dine with him, and he with me."*
>
> Revelation 3 verse 20

Again I ask, are we good Christians? Are we '<u>Christ-in</u>' people?

I came across a poem in the early days of drafting this book. At first I thought to include it here (subject to © permission). But then it seemed that to include it in its entirety was not required. The poem was, "If Jesus Came To Your House Today", by Lois Blanchard Eades. I first heard it when playing a copy of one of J. John's speeches made in the series, 'Battle 4 Britain', and then read various versions when I researched it on the Internet.

The theme of that poem is of wondering what we would do or behave like if Jesus did, in fact, call on us. But, although looking at that could well be thought provoking, I feel that as Christians we should be doing more than considering His *visiting* us. Tidying up and hiding certain items in case He calls, indeed! Does not inviting Jesus into our lives require even greater commitment? What are we, part-time Christians? Does He not see all anyway? Was our commitment just so that we could be saved?

For a long time I have observed and been led to wonder at the quality of our Christian Commitment to the Lord and each other. Perhaps there is just misunderstanding of what the Lord requires of us - or am I being too generous! Perhaps it is because the enemy has infiltrated our thoughts.[3] We can get sidetracked, then lose out on the relationship with God, which He desires, and the perfect plan He has for our lives.

What makes me say this? For many years I just felt that much more should be happening within the church. I have been concerned by our way of life, not least my own! I have also been taken aback by the replies to promptings and encouragement to aim for a deeper commitment. I have heard the responses, "You only want to see the spectacular", and, "There are no second class Christians!" Of course, in the eyes of God, we **are** all equal. But did the latter response suggest that **all** Christians **are** doing well? Surely all

---

[3] I believe that if we allow the enemy to get established in our thoughts; we can be diverted from God's perfect plan for our lives. I have been prompted to research this matter further, and record it in part two, 'From Deception to Deliverance.'

Christians are not doing a 'first class' job? Would you not agree that the behaviour of Christians is not always exemplary? To our shame, I feel that many of us are falling short of what our Lord requires of us. It should not be that people are put off becoming Christians and miss salvation because we lack sensitivity, or set a bad example, or lack real commitment; or indeed, act in any way contrary to the love of Jesus. Thank God that there are very loving and devoted Christians in whom God is obviously present by His Holy Spirit. Thank God for their commitment and example. Where would we be without them?

Becoming a Christian is most important. It means for each of us that Jesus did not die in vain! It also assures us of our eternal destiny, and ensures us a blessed life whilst waiting - when we walk close to God. By walking closer to God, we can become better witnesses, others can better be encouraged to become Christians, and yet others be drawn closer to God themselves. We are blessed by finding within ourselves a peace and joy hitherto unknown! That also draws others unto the Lord!

It is to encourage us to walk closer to God – to accept all He has for us, and for us to let Him work through us - that the following chapters are written.

Chapter Two

BEING FILLED WITH THE HOLY SPIRIT

As chapter one suggests, there is more need. It offers a challenge because many of us are falling short of a fulfilled Christian living and witness. Though we may try our best, there seems to be something missing. That's just what it is! There is something missing! We have been trying to do **our** best when we should have been succeeding with the empowering of the Holy Spirit.

I would like to expand on what I have just said above, 'though we may try our best'. There are Christians who do all they can to learn from the Word of God; they spend hours and hours studying the Bible. Some have attended many study groups, and have memorised many scriptures. I have known leaders of some study groups to spend many hours in effort and research. Yet, for all the above, if we do not put our learning into practice, have we not just had an academic exercise? You have no doubt heard the saying, 'action speaks louder than words'. Are our minds just filled with many words - read or spoken? Are we any nearer being in the kingdom of God?

I think these scriptures prompt us to consider that there is more than just learning the Word:

> *'For the kingdom of God is not in word but in power.'*
>
> 1 Corinthians 4 verse 20

> *'But be doers of the word, and not hearers only, deceiving yourselves'.*
>
> James 1 verse 22

Yes, studying the Word of God is important and has great worth, but we should take note that the kingdom of God requires power.

Let me give a modern day analogy.

We would not dream of getting into a car to go on a journey without making sure there was enough fuel in the gas tank. Or would we? I know I did on one occasion! My mind was on other things and I was hasty to get underway. Shortly after setting out, the car lost power and came to a stop. It was out of fuel! We do not get far with insufficient fuel! To travel with the car then is so much harder - it takes so much longer to *push* without fuel! I needed fuel to journey on. So it is with the Christian journey - **we** need fuel. We need to be filled with the Holy Spirit! And we need a good driver – **guess who**!

> *"So I say to you, ask, and it will be given to you; seek, and you will find; knock, and it will be opened to you.*
>
> *For everyone who asks receives, and to he who seeks finds, and to him who knocks it will be opened.*
>
> *If a son asks for bread from any father among you, will he give him a stone? Or if he asks for a fish, will he give him a serpent instead of a fish?*
>
> *Or if he asks for an egg, will he offer him a scorpion?*
>
> *If you then, being evil, know how to give good gifts to your children, how much more will your heavenly Father give the Holy Spirit to those who ask of Him!"*
>
> Luke 11 verses 9 to 13

Yes, our heavenly Father actually desires to give us the Holy Spirit!

Now the gifts of the Holy Spirit are these:

- The word of wisdom
- The word of knowledge
- Faith
- Gifts of healings
- Working of miracles
- Prophesy
- Discerning of spirits
- Different kinds of tongues
- Interpretation of tongues

See 1 Corinthians 12 verses 8 to 10

If we continue reading on from 1 Corinthians 12 verse 10, we see that He distributes to each one individually as He wills. So no one gets it all! We are told in scripture that the distribution is for the body (of believers). It does say that we do need one another! However, although one may have a particular gift (or even more than one) as a ministry, I believe that the Holy Spirit can manifest any gift in any one of us should the occasion warrant. Example - when no other Christian is available at a particular time and it should be necessary. But He will need our willingness, our co-operation. The Holy Spirit is a gentleman. We still have free will, and He will not force us against that. But isn't this exciting? To think that God would bless us and use us this way...

And the fruit of the Holy Spirit is:
- Love
- Joy
- Peace
- Longsuffering = patience
- Kindness
- Goodness
- Faithfulness
- Gentleness
- Self control

See Galatians 5 verses 22 & 23

Whereas we may individually have only some of the *gifts* of the Holy Spirit, we should have **all** of the *fruit* of the Holy Spirit. (The word is fruit not fruits). Now isn't that interesting - and exciting too? Can any loving person not *delight* in the manifestation of those qualities? Indeed, do we not *desire* such qualities? And does the scripture not say:

> *'Delight yourself also in the Lord,*
> *And He shall give you the desires of your*
> *heart.'*
>
> Psalm 37 verses 4

I don't know about you, but I say, "Yes please. Fill me with the Holy Spirit."

Chapter Three

CLAIMING THE PROMISES OF GOD

In the last chapter we quoted a promise of God in Psalm 37.

'Delight yourself in the Lord,
And He shall give you the desires of your heart.'

Is it not amazing that God who created the heavens gives **us** (mere mortals), a promise? In fact He gives us many promises!

However, have you noticed something?

Almost always there are **conditions**!

If we fulfil these conditions, our heavenly Father will keep His promises! Any doubt comes from the enemy. God cannot lie!

So let us meet the conditions, and receive from God what He desires us to have.

I would like to encourage you to research each

promise God puts on your heart. Fulfil each condition with God's grace. Then be confident that God will, because of His word and His love for you, keep the promise.

There it is! No but's; no if's!

Surely I need say no more on this subject?

Chapter Four
DOING WHAT GOD ASKS

Of course, in The Old Testament, God gave Moses the Ten Commandments for people to live a righteous life. I am sure that I do not have to elucidate on these as recorded in Exodus 20. However, if a study were wanted, I would highly recommend what J. John says on these Ten Commandments.

Now, to encourage our confidence to do what God asks, I would like to look at the authority God gives us in The New Testament.

> *'Then He called His twelve disciples together and gave them power and authority over all demons, and to cure diseases.*
>
> *He sent them to preach the kingdom of God and heal the sick.'*
>
> Luke 9 verses 1 & 2

'After these things the Lord appointed seventy others also, and sent them two by two before His face into every city and place where He Himself was about to go.'

Luke 10 verse 1

'Then the seventy returned with joy, saying, "Lord, even the demons are subject to us in Your name."

Luke 10 verse 17

'And He said to them, "Go into all the world and preach the gospel to every creature.

He who believes and is baptised will be saved; but he who does not believe will be condemned.

And these signs will follow those who believe: In My name they will cast out demons; they will speak with new tongues; they will take up serpents; and if they drink anything deadly, it will by no means hurt them; they will lay hands on the sick, and they will recover."'

Mark 16 verses 15 to 18

I think it is clear in the gospel of Luke, that disciples were to:
- Preach the gospel.
- Heal the sick.

- Cast out demons.

And in the gospel of Mark, '*Those who believe*', will be able to do the same. '**Those who believe.**' We're part of that; that's **us**!

Yet what are we doing about it?

Regarding the above instructions:

It is noted that there are many who:

- Preach the gospel.

There is some movement to:

- Heal the sick.

But few:

- Cast out demons?

Or even seem to know anything about the subject!

You only have to consider the state of the world to realise the enemy is getting away with an awful lot!!!

I know there are many mighty servants of God throughout the world who are channels for God's blessings. Miracles are happening; there are many healings; some are being freed from demonic activity. In some geographical locations, it is as if The Acts of The Apostles were continuing. But could not a few of us do more than we are?

You may feel the Lord only blesses a few with the responsibility of 'chosen' work. If so, why? Yes, the Word does say:

"For many are called, but few are chosen."[4]
Matthew 22 verse 14

So what has happened to 'the many who are called'? Well, like in the parable that precedes this verse, many do not respond the correct way to what the Lord wishes from them! Like so many today, many do not want the kingdom of heaven.

NO! It is no good saying that only some were called and others were not. It is no good saying that some are called but we are not. I feel God is assuring me that we cannot say we are not called! That is the devil's ploy to deceive[5] us into being inoperative.

To back this up, let us remember that many early manuscripts from which the Bible was translated were in Greek.

In Greek you might say, 'The many are on time, but the few are late.' The English equivalent is, 'Most are on time, but some are late.' In Greek, 'the many' and 'the few' add up to everyone; just as in English, 'most' and 'some' add up to everyone. Still not sure? How about the example - most have voted, but some have

---

[4] It is interesting that this verse comes at the end of Jesus' parable starting chapter 22 verse 1. The thirteen verses spoke about the kingdom of heaven.
[5] There's deception again!

not? That adds up to all.

Some of you at this point will probably say, "What about the millions who have not heard the message of salvation; are they not also part of the called?" Yes, they are. God is calling them too! Is this not why he tells His disciples:

"Go into all the world and preach the gospel to every creature.

He who believes and is baptized will be saved; but he who does not believe will be condemned."
Mark 16 verses 15 & 16

I trust we have established the fact that everyone is called. So what is our part in all this? To what are we called?

Firstly let me say that we may be called to the other side of the world; or we may be required to witness to the person next to us. But be not concerned; God knows exactly what is best - and will give grace accordingly. He will not ask us to do something we are not able to do. It will be His ability; we just have to offer our willingness, and our availability.

But wherever we may be needed, or for whatever we may be needed, it is most important to remember we are the Lord's ambassadors. It is fitting, therefore, that we show others that we have something more than the non-Christian. We should, of course, exhibit the fruit of the Holy Spirit.

I am going to put forward the challenge that each one of us asks God what He wants from us - over and above living a righteous[6] life, that is. However, let Him confirm any special direction or work through other Spirit filled Christians.

I think this confirmation from others is necessary, because I have witnessed some situations, which have had the alarm bells ringing. I mention two observations. The first was someone who believed he was being directed by God but would later listen to no one else. He got himself into a right pickle and was admitted into an institution. The second is someone who thinks he is gifted in prophecy, but the words seem to come out with such harshness! It makes one wonder if he is trying to impress others by implying that he has a direct 'telephone line to God'.

For those who believe they have a direct 'telephone line to God', I would ask these things. Is each message 'laid before the Lord'? Do others agree any message is of the Lord? Does scripture agree? If other Spirit filled Christians cannot agree, is it not really suspect?

---

[6] Righteousness. See 1 Corinthians chapter 6 verses 9 & 10.

'For God is not an author of confusion but of peace, as in all the churches of the saints.'
1 Corinthians 14 verse 33

So:

'Beloved, do not believe every spirit, but test the spirits, whether they are of God; because many false prophets have gone out into the world'.[7]
1 John 4 verse 1

Therefore:

'Do not quench the Spirit.[8]
Do not despise prophecies.
Test all things; hold fast what is good.
Abstain from every form of evil.'
1 Thessalonians 5 verses 19 to 22

To sum up this section:

It is important to do what God wants. In our enthusiasm, however, it is still possible to be open to deception. The enemy will try to make us believe we are God's gift to mankind. Only Jesus could fulfil that role! We must be on our guard against pride. To counter pride we ought to be humble. Please remember Jesus taught

---

[7] Author's note: They probably thought they were doing the right thing.
[8] Note capital 'S' i.e. The Holy Spirit.

this when He washed the disciple's feet. The hymn 'Amazing Grace' is for me here. I feel it is amazing that a wretch like me is writing this. Not only was I in my twenties before finally passing my English language G.C.E. exam, but also I have been a slow learner in other areas. St. Paul said that he was the chief of sinners; but was in ignorance. I have done things, which he would not have done; but in some I was not ignorant! I ask here for forgiveness from any I have offended in any way. Please forgive me.

Chapter Five

COMFORTING OTHERS

'"Comfort, yes, Comfort My people!" Says your God.'

Isaiah 40 verse 1

So I wonder what our reaction to such as the following poem is:

'I was hungry'

'I was hungry.
And you formed a humanities club and discussed my hunger.

I was imprisoned.
And you crept of quietly to your church and prayed for my release.

I was naked.
And in your mind you debated the morality of my appearance.

I was sick.

And you knelt and thanked God for your health.

I was homeless.

And you preached to me of the spiritual shelter of the love of God.

I was lonely.

And you left me alone to pray for me.

You seemed so holy, so close to God.

But I am still very hungry, and lonely, and cold.'

          Anon.

My questions should surely be obvious. "Do you think that person was sufficiently comforted?" "Did the church fulfil its duty?" "Was God pleased with the church?"

The above poem undoubtedly was based on Matthew 25 verses 42&43. It may be that the writer (of the poem) was bitter because of being let down by the church over some issue. It may be that he or she was not even a Christian. But, whatever the reason, this should not to be taken lightly. It is noteworthy how Jesus responds to such reactions to cries for help:

> *"Assuredly, I say to you, inasmuch as you did not do it to one of the least of these, you did not do it to me.*
>
> *And these will go away into everlasting*

punishment, but the righteous into eternal life."
Matthew 25 Verses 45 & 46

It would very much seem that comforting the needy should be something more than a soft option! There is real need for us to understand, by the above, how serious God considers our lack of help or comfort to others when it is required!

Of course, we remember the Lord giving the story of 'the good Samaritan'. What do you think the one who had been robbed thought about the church? What do you think about Christians who respond in the same way? Well, let's not judge; rather let us remember, 'There but for the grace of God go I', and ask for compassion and wisdom as to what to do in situations as they arise.

At the risk of labouring this matter, I offer these three examples:

I make mention of two separate occasions where someone was with Christians before committing suicide. In each the Christians have stated that they could do no more!

Thirdly, many years ago, I personally was overwhelmed with someone's behaviour and felt I needed help. I asked an elder of a church for the elders to help in prayer. Was I wrong to be shocked by the reply, "He does not belong to our fold." Are we not all brothers and sisters in Christ? Have we not all experienced disappointments to cries for help? Was not even our Lord disappointed when, in the garden of Gethsemane, He asked for the disciples to watch and pray?

In this present age it is a fact that many Christians give the impression that all they needed to do was to be saved. Others attend church only for christenings, marriages and funerals. Yet others will attend church regularly almost as a sense of duty - but even that can change if a new church leader is appointed, or someone takes their regular seat, or the service time is changed! The idea that each Christian has a job to do seems to elude many. How sad! It is sad that not only God's kingdom loses out! Yes, that's right, the ones who are not living a Christ centred life are also certainly losing out! I guess they just do not know the joy and, sometimes, the sheer thrill of walking in Christ's footsteps. Some say, "But if I do 'church' things I may not have enough time for my other planned activities." Well, many others say that they have made 'putting God first' their priority. They have made time to be a 'Good Samaritan'. What do they report - "Somehow there **was** enough time and enjoyment as well."

Yes, God is no man's debtor!

In the Beatitudes we read:

> *"Therefore be merciful, just as your Father also is merciful."*
>
> Luke 6 verse 36

And, later:

> *"Give, and it will be given you: good measure, pressed down, shaken together, and running over will be put into your bosom. For with the same measure that you use, it will be measured back to you."*
>
> Luke 6 verse 38

But, even before that scripture, way back in The Old Testament:

> *'He who has pity on the poor lends to the Lord, And He will pay back what he has given.'*
>
> Proverbs 19 verse 17

And, returning to The New Testament:

> *'But this I say: He who sows sparingly will also reap sparingly, and he who sows bountifully will also reap bountifully.*
>
> *So let each one give as he purposes in his heart, not grudgingly or of necessity; for God loves a*

cheerful giver.

And God is able to make all grace abound towards you, that you, always having all sufficiency in all things, may have an abundance for every good work.'
   2 Corinthians 9 verses 6 to 8

But surely we do not do it knowing that we have a reward for so doing?

I remember a time when I stayed in Canada for a while just after I came out of teaching. My brother (he had emigrated some years before), and family had some work to do on an island, and I stayed in Victoria. A friend lent me a house to stay in. Yes, a whole house! When I asked how much I could give, I was told, "Just pass it on to another." What is particularly noteworthy about this incident is that that person was not a Christian! Surely we can do as much? Well, perhaps not lend a whole house, but you know what I mean!

To a greater or lesser extent we all have needs and problems, and are under attack one way or another. We should be sensitive about one another's situations. We should take responsibility to encourage, comfort, and pray for each other!

Chapter Six

PRAYER

'Take responsibility to encourage, comfort, and pray for each other' was mentioned in the last chapter.

Should I have placed prayer first? Some may feel that comforting can be natural - perhaps without even thinking, whereas they may think that prayer has to be thought about before offering it up. For example, every normal mother instinctively comforts a needy baby. So why should prayer be suggested first? Would a mother hearing a crying child first think, "I must pray about this", then rush to the fallen child? You may well say that God has given us common sense, and that the suggestion above is 'rather rich'. And on the face of it I would agree! But I am looking at prayer in a much deeper way; a way, which I hope, will be challenging, and will show a whole new dimension to prayer for some readers.

Before we come to this new dimension, let us not leave the matter of the crying child 'in the air'. Poor mum! There are two possibilities here for not praying last: one is to offer up an arrow prayer as one sets off simultaneously[9] to investigate the cause of crying; the other is to set off knowing that right from birth (most probably since pregnancy was discovered), a prayer had **already** been made to cover the child's ongoing safety, and another prayer for wisdom in handling each other situation as it arises. (Missed that one? Better do it right now.) Forgive me if you have those covered and feel I need not mention what you have thought obvious, but if just one child benefits from such a prayer, missed but now rectified, then this reminder was necessary.

'Should prayer be first', was the question earlier. How many of us have come up against a problem and tried all ways to resolve it *'in our own strength',* and then, having failed, turned to prayer as a last resort? Is it pride that prevents us asking God first? Perhaps we don't wish to bother Him with what we deem an insignificant problem. I am sure there are many reasons, but the saddest must be that some just don't believe

---

[9] When I hear the siren of an emergency vehicle, I immediately pray for the situation, which has been 'advertised' by the sound. (Even if driving, an arrow prayer of, "God bless", at least can be asked).

He cares enough about us to answer. Now, where do you think that thought came from? But scripture records Christ's own words:

"Are not two sparrows sold for a copper coin? And not one of them falls to the ground apart from your Father's will.

But the very hairs on your head are all numbered.

Do not fear therefore; you are more value than many sparrows."

Matthew 10 verses 29 to 31

Yes, He cares for us! But He will not treat us as robots. We have been given free will and He will not overrule our decisions - even if He disagrees with them! Mind you, we may well come to regret some of our decisions! He will, however, bless us if we put matters into His hands. We have to mean, "Thy will be done", as in the Lord's Prayer - (the perfect prayer containing all needs). Are we prepared to accept His decisions? After all, they may be "Yes" or "Wait" or "No". But whatever the outcome, it will be for the best. We need to trust that He knows best!

Trust. That reminds me of a wonderful encounter. Some years ago I was blessed personally with the following experience:

I was walking in a wood some half-mile or more from a farm when I heard a sheep bleating outside the wire fence. Inside the fence its lamb was also distressed. It had found a hole large enough to get through but now could not find the way back. What was I to do? The time it would take to get to the farm to report it - even if there was anyone there - may be too long and too late for the lamb. There were foxes and other dangers in the wood! I was reminded of the healing I had had in the name of Jesus. The thought came, "Why should faith only benefit humans?" So I called out, "In the name of Jesus, come here." The lamb ran to me and stood between my boots. I bent down and picked up the lamb astonished that it did not even struggle. I went to the fence and lowered the lamb over the wire. It is almost indescribable the heartfelt joy I witnessed as the lamb was safely re-united to the ewe.

Immediately came the thought, "Jesus is the Shepherd; we are the sheep; we need to trust."

I thank God so much for that experience.

I remember another experience:

Many years ago I was driving an estate car on a narrow country road, when I had to stop behind a bus that had come to a halt for oncoming traffic. The driver decided to reverse.

I tooted my horn, but the bus kept on backing - and backing. The front end of the car sank as the luggage compartment of the bus engulfed it. Now I didn't wish to travel **with** the bus, so we parted the two vehicles. My bonnet was still somewhat lowered - but the car still ran and steered OK. Now that car had a tailgate that hinged down, not up - so, when it rained, water would get in and fill the well in which the spare wheel normally rested. You could not drill a hole through the metal to let the water out - for it was the roof of the petrol tank! Third problem - the clutch had started to slip. Struggling up a hill some time later, I blurted out, "God, how much longer do I have to put up with this car?" Was that a prayer? Anyway, God heard it! Silly me! Now, seasoned Christians should know what followed! You've got it! I only had that car in a (somewhat) roadworthy condition for about another hundred yards! It was then not roadworthy at all! If there had been a frost, the road for three miles earlier would normally have shown signs of ice. Not a trace. But now! And such a little patch! (I was to discover when I got out). That was after I lowered myself down from the upside-down position and squeezed out of the window! On the slightest of bends, and not at a great speed (remember, the clutch was slipping), the rear end of the car had decided to overtake the front end. It clipped the kerb, and the car flipped over! Now it was not just the bonnet that was rather squashed! Definitely I did not have to 'put up with that car' - it was now a write-off! Yet I was not injured[10] - not even by the spare tyre, which (because of the puddle in the back) was loose on the passenger seat. It could have fallen on me as we turned over! The thermos flask of coffee, which had been resting by the spare tyre, was also whole. By the way, I was a teacher at that time - and this incident happened perhaps only two hundred yards from the secondary school I taught at. You've got it again! Some pupils **had** witnessed the incident - as I found out in more than one lesson that day!

Now, I ask you, does God have a sense of humour or not? And, another thing, He doesn't always answer prayers the way we expect!!!

Not convinced? Let me report an incident told me by a brother:

His car had finished in the ditch by the side of the road. (This was in the days before mobile phones etc.) Such was the position that logic seemed to suggest that it would probably take a

---

[10] One of the many, many times I believe divine intervention prevented a serious injury in my life. I record some others in chapter seven.

dozen people to manhandle it out. So, what did he ask the Lord for? He asked for a minibus of twelve people to come by. The first vehicle was a car, which stopped, and the occupant offered help. He responded with, "Thanks for offering, but I've asked the Lord for twelve people to come by." Well, he did ask! He says that the car sped away **at great speed!** Eventually a minibus did come by - and it did have twelve people in it! However they could not help - they were rugby players, and had celebrated. They were now rather the worse for the liquid refreshment they had consumed, and incapable of helping!

He tells me he then got a lift and arranged for a garage to collect the car the next day.

I do not know more. Perhaps God knew something was wrong under the car. If that was the case, had it been manhandled, it could have been dangerous to move or drive without the garage first checking its roadworthiness.

But not all communications with our heavenly Father are as noteworthy as the above examples. But, patience please, your turn may yet come…

There are some who trust God to answer prayers, but treat Him as a daddy who keeps giving out presents. They give Him a 'shopping list' - and expect Him to provide. Yes, God does like to give us good gifts - but not just presents like Santa gives children at Christmas. {Oh, you don't believe in Santa? Well, to our shame, there are some who don't believe in God!} He desires to present Himself likened to our best companion - and then some! No one could have our best interests more at heart! Indeed, He wants us to cast our cares upon Him, talk to Him, **and listen to Him**! Remember in an earlier chapter we considered 'Waiting on the Lord'? Is this not part of the whole new dimension to prayer that I promised earlier? Well, practice makes perfect. What a challenge! For those who have not yet been that confident in God's love, and the relationship He wants to have with you, you're in for a surprise - and a very blessed one at that!

If you take this to heart, prayer will never be the same again!

Chapter Seven
GOD'S PROTECTION

Earlier I was writing about one of the accidents I have had in cars. You may remember that the car was upside down and a write off, but I had no injuries. I believed that I had been being blessed with divine intervention. If you don't believe the same, by the time I have finished this chapter, you sure won't wish to travel with me! Already you are no doubt wondering about the above, 'One of the accidents'. "**One**? How many has the chap had?" If you are an insurance agent, please do not read on[11]. On the other hand, by the time you have read on, you may well believe in divine intervention, and the generosity of God. Such being the case, having had so many scrapes with serious injury or worse, my experiences are of some worth, and testify to God's protection.

You asked, "How many has the chap had?" That was, of course, just referring to cars, was it

---

[11] Spiritually speaking, I don't mean that!

not? The answer is eight. But, if you meant road accidents, then there were two motorcycles as well. I am not going to go into the other car accidents, as they are somewhat boring, but I will the motorcycles.

One - a 350cc Triumph decided to break a con rod at 53 mph. I was happily going along when there was a bang and loss of power. I drew up, looked down, and noticed the top half of the engine had parted company with the rest. It was dripping oil over my boot. Had I been going slower, would it have jammed and thrown me?

My second bike was a 250 BSA. One day I was overtaking a line of cars. Coming the other way, there was a lorry overtaking a cyclist. But there was **still** plenty of room. Then a car decided to overtake them both. Now there was **not** enough room! I do not remember the impact. I came to at the side of the road holding a blood stained nappy or towel to my eye, (the goggles were smashed, and glass had got into my right eye),[12] I had a broken finger, but I could have been far worse off. This was in the 1950's when crash helmets were not compulsory. I wore one - I

---

[12] Two weeks after the accident a flake of glass came out from my eye. However, for years afterwards I was able to win trophies for shooting – one of my main hobbies. I mention this; not in pride but to show the extent of how good was my eyesight still. Yes, I did use my right eye. My thanks are to God.

would not be writing this had I not. There was now a large hole in it! My mother was working in an office at the time. A colleague came in and reported that he had just hit a motorcyclist. A few moments later my mother had a phone call from the hospital! My poor mother – she had so much to put up with. I am sure she is more at rest with the Lord now.

Much earlier in my life there were other scrapes with death. Once we (I have two younger brothers) found some plastic cord-like material in a hedge by a US Army installation. Being of the usual inquisitive sex of our species, we discovered that it sparkled as we plucked off lengths and lit it. Great! We took a whole bundle home. We decided to continue our observation by the fireplace in our living room. Wow! Indoor fireworks! Our father came, but didn't say a word - he just snatched up the bundle and went to the bottom of our garden. He proceeded to lay a trail of some pieces - and lit the end. The neighbour's window across other gardens opened to find out what the WOOMPH was. It was, in fact, cordite[13] exploding into a ball of fire. Now, even gunpowder just flares if enough air is around it - but not in an enclosed space! Our

---

[13] Cordite was used as a charge in anti aircraft shells during the last war.

living room was an enclosed space! And still was when we last looked at our old house some fifty years later - I am sure that would not have been the case if… At least, not without rebuilding!

I visited that same area when I was a little older. Some chaps were drilling the concrete bases which had supported the anti aircraft guns during the war. It was open ground next to a cemetery. They asked me to guard the back exit gate to the cemetery, and turn anyone back. No one came that way – much to my disappointment – I wanted to see the expression on his or her face as I passed on that 'request'. You see I had an air rifle crooked over my arm at the time. The cemetery closed and the workmen called me over to show me what they had been doing. The timing was just right. I saw them putting a tube of something into the hole on a 'string'. Well, it wasn't string; it was a fuse - which one of them lit. "Quick, run!" We sheltered behind a shallow ridge. Bang! Weeeee! A chunk of concrete whizzed over our heads - and embedded itself into the front end of the compressor used for the pneumatic drill! I was told it was just a half-pound of gelignite[14]. The concrete, which had been shattered, was four foot six inches in depth!

---

[14] Another explosive.

The next few years were fairly uneventful until I was in my late teens. Well, that is if you don't count making home made powders for cap bombs from chemicals, which can no longer be bought from chemists. I wonder why? Mind you, some poor boys have been badly injured using the same. So it really is beyond a joke.

So what happened when I was in my late teens? Three instances stand out.

- ### Instance number one.

Again concerned with explosives. I was a cadet in the Air Training Corps. At one airfield a few of us were invited into an aircraft, which was to do a practice-bombing run off the coast. We took off, flew a while and had a few splashes near the floating triangles. We then returned prematurely to the airfield. That was the smoothest landing ever. I did not feel a thing. As I looked back after alighting, some airmen were removing a 25lb bomb which had become wedged!

- ### Instance number two.

Whilst still a cadet. I had been learning how to fly a glider, and it was time for my solos. I took off alone (one does on a solo), and climbed to 900 feet. Nose forward, then cast off from the winch. That, I **had** been trained to do. But, **EEK, I WAS ON MY OWN!** I let go of the joystick and grabbed the sides of the cockpit. Now, that is something you are **not** trained to do! As you know there is no engine in a glider. What did I have - a few seconds at the most before it stalled or went into a dive? 900 feet! But enough time for God to make me get a grip - of myself - and the joystick. But it didn't finish there. In more ways than one! On my third solo (the instructor did not know what had happened on my first), I was asked to climb as high as the winch would let me, and do a complete circuit. As I climbed, I calmly just looked at the altimeter and felt for the change of pull from the winch. This would tell me I could get no higher. When that was so, I cast off. 1,350 feet. **Now** I decided to look down. Nothing! Last flight on a December afternoon as dusk was falling. A little later I could just make out a coloured rectangle. Good – the winch. But it was not! Now, at 700 feet, I discovered it was the canvas roof of a vehicle - outside the mess. Now this place where we ate meals was not by the grass landing area. In fact it was some distance away - over buildings. My circuit was cut short as I cut corners to make the landing area. There was quite a bump, which caused the others to rush over. Their remark, "Is the glider alright?"

    I did not glide for over forty years. Then I

went to New Zealand to visit my eldest daughter and granddaughter. I came across some Para-gliders at a tourist centre, and was blessed with this experience. This time it was in tandem! On the way down, I asked if a Para-glider could do aerobatics. I was **only** asking! The pilot **showed** me it **could**! I guess we can still be 'adventurous' even if we are in our late sixties! But is not life one long adventure? I thank God that I have discovered how much adventure there can be in the Christian walk. But, for now, let's get back to the present subject:

- **Instance number three.**

I had been nominated, by the A.T.C. to go on a months Outward Bound Course. This one was at a mountain school in the Lake District. I had never set foot on a mountain before. When a manager from where I was working heard about the course, he offered to 'introduce' me to the mountains. He was taking some scouts to Snowdonia. After the car journey, he suggested that we refresh ourselves in a lake (Llyn Ogwen) right next to where the tent was pitched.

Now, I was never really keen on swimming. Something about being pushed in before I can remember! But mother got me to have lessons. I do remember being pulled by a rope from the side of a swimming pool. Then once I was asked to go the other way. I was half way across before I realised that the rope was not pulling me: I was pulling it! Glug! The instructor said something about, "Well you can swim now." So, years later, now at Ogwen I thought just a little distance would be OK. I went to turn around and… I sank! I came up three times; I went down three times – and stayed down! By now the others had climbed out, and buried their heads into the tent to sort out some things. My predicament went unobserved. But now, like in the glider, a sense of calm came over me (I prefer to believe God gave me a sense of calm each time). I noticed the rocks sloped upwards. I simply had to clamber over them until my head was above water. I could breathe again!

The next day I climbed my first mountain, Tryfan, a climb of 2000 feet. That night it poured and poured. The next morning we abandoned the expedition and headed home. I think we were one of the last through the pass. We heard later there had been a landslide, which blocked the pass.

After that the Outward Bound Course was a great experience.

Just a year or two after attending the Outward Bound Course, I joined the Royal Air Force. After the first year I was posted to Cyprus. As a result of the EOKA terrorists we could not leave camp for recreation - and work finished at lunchtime! Then I noticed great glee coming from vehicles travelling on a Thursday afternoon and at weekends. It was the Nicosia Mountain Rescue Team going to and returning from trips to the mountains. What a happy bunch they were! Would I with just one day and one months experience in the hills have a chance of joining them? I guess God wanted that as part of my experiences. What a great time we had on our trips. I could tell many tales of my two years in Cyprus – both on duty and with the Team. Many could come under this chapter's title. But I am going to mention just one.

I was almost twenty-one years of age, and I had a birthday card from my cousin in America. One of the things different about this card was that it was dropped to me by parachute in Turkey. We were at 14,000 feet in early May, and on a rescue mission. At least that was the reason we were given at the time. The mountain in eastern Turkey, some 400 miles from the nearest airfield, seemed very cold – especially after we were used to the temperatures in Cyprus. The party that went up to search for survivors came back unsuccessfully – and were pretty worn out. I think some had frostbite. Then the orders came through, "Destroy the cargo at all costs." Now when the military are given **that** order, it means just **that**! The cargo was rockets bound for testing in Australia. Now they were strewn across the mountaintop, which was only 130 miles from Russia - and it was in the days of the cold war! Five of us had not been to the plane crash site. We were led by George who had been in the first party, but was now the only one who was fit enough to return. (A great chap George. One well worthy of the British Empire Medal he received for leading us. We had many adventures together. I hope we meet up again one day.) The six of us were dropped off one at a time by helicopter at 12,000 feet. It was a tense time waiting for the others to join - especially when a Kurd, rifle slung over his shoulder, came down a ridge. We thought he had been up to the crash.

When we arrived on top of the mountain I switched on the radio,[15] immediately the valves blew. Yes, it was cold up there! Now we had no communication with base! After some time an aircraft came through the clouds with its bomb doors open. Now hadn't it been said, "Destroy this cargo at all costs!" No communication with base! I fired the Very pistol to show that we

---

[15] This was in the days before mobile phones.

were there, and where we were. I had allowed for wind[16]. The pilot later really shook my hand, and remarked, "You sure made us see you - you just missed the cockpit!" The plane veered off, came back, and dropped parachutes, the canisters under which contained blankets, food, tents and explosives to shape the rocket cases beyond recognition. No, this time I did not get to play. But it was no play for the two that did that work - they were so tired after escaping many explosions. They took a half hour to walk the last hundred yards into camp! No going down that day!

Camp? Well, we had two two-man tents. So the **six** of us should have been warm, yes? No! Our boots were frozen and our breath condensed on the inside of the canvass, which then shimmered with ice crystals. Pretty? Pretty cold! The tents were tied together at the entrances with the front guy ropes. There was no more than three feet between them. In that space the canned food was stored. In the night I heard some of them move. I remember calling out, "Is that you George?" No answer. Next morning

---

[16] A term used in rifle shooting. We sometimes aim off for a strong cross wind because the bullet is blown off course. Whilst shooting at targets five hundred yards away, I have even aimed at the next target to hit my own. And yes, I did hit my own - before you ask!

we observed bear tracks between the two tents. We were told later that only **small** bears were in that part of Turkey.

That morning we set off down. After a short way we had to stop. There was a whiteout.[17] I could actually see my hand vanish as I moved it away from my face! Our feet were further away! We dare not set off down in these conditions. We dug a snow hole and got in. I just wanted to sleep. Maybe I would not wake up, but I was so, so tired. Once again I give credit to another. This time George was His instrument. Whether it was George's singing or his using my head as a drum, I do not know; but it **did** keep me awake! The whiteout eventually passed and we set off again. A step or two on top of the snow, then up to the waist as one sank! Sometimes just one leg would go through the crust. So tiring - but we had the satisfaction of knowing a job well done, and we were on our way down.

We came out of the cloud and received a great shock. The whole of base camp seemed lined up ready for the off. It transpired that the day before they had heard just one bang. They thought we had blown everything up - yes, including ourselves. They could do nothing else. No

---

[17] A condition where ice crystals in the air, cloud, and sun behind, give conditions far worse than the thickest fog.

one was fit enough to return to the mountain, and even if they had been, the conditions had deteriorated dramatically! We had just one signal left - a red smoke grenade. I let it off. Green smoke suddenly drifted near the convoy below! A reply! We had been seen! Soon a helicopter started up and climbed to our location. We were airlifted in two's - down to safety. Had the red smoke been covered by drifting cloud, or had it not been seen, or... To be left alone in that area? I do not think we had much chance of surviving. I will not say why, but our Turkish Army escort apparently had shot two Kurds! I hesitate to think our fate for the six of us at the hands of the Kurds under the circumstances!

There have been many instances of brushes with disaster in the forty years since my service in Cyprus, including falling out of a tree, my head being first to hit - a rock. So that's why I am as I am, you say? Well, maybe it did knock some sense into me. Well, someone had to!

I think the point has been made. I believe there is enough evidence of divine intervention. I really do think that without God's protection I would not be writing this.

Why? Perhaps the answer lies in the epilogue to this book.

Chapter eight

BEING A DISCIPLE

We have looked at what really is a Christian, being filled with the Holy Spirit, claiming the promises of God, doing what God asks comforting others and prayer. I have included a chapter giving some of my experiences to show just how much God's intervention can protect one of His children. Life as a Christian can be really amazing.

You may not have had as much excitement as I have recorded; or you may have had more. Nevertheless you will be somewhere on that Christian journey. You will be somewhere between the beginning and the end.

Jesus is, of course, the Alpha and the Omega - as in the Greek alphabet - the beginning and the end. What I have tried to convey is that becoming a Christian should have been just the beginning of our life in Jesus Christ. I suggest the 'Alpha' of our life in the Lord.

There has been tremendous effort and response to 'Alpha Courses', and many thousands (perhaps millions) have come to know the Lord and been saved as a result. There are follow up courses where Christians can be strengthened and advanced. After that, apart from attending church services, one can go to midweek meetings (though not necessarily 'midweek'). Some are called Bible-Study groups. Some are called 'Share and Care' groups. There are groups that gather for prayer. Many present day Christians have matured and have been called to do great works. Missionaries are still allowed in many countries and are spreading the Word of God. Miracles are still happening. There are countless healings. Evangelists are being used to convert millions via the media of television in over a hundred countries.

So, can anything else be done?

Let me ask, "Do you think God is happy with all of us?"

Whilst blessed Christians are ministering to millions, many millions are still unsaved. It is for the salvation of these, and for the rest of us to progress in doing all God asks (for He alone knows what is needed), that I suggest some of us could be doing more. Are we prepared to be used to the Glory of God? Maybe many of us will not be called to evangelise great crowds, or be a channel of some miracle, but, and here is the crunch, – each of us can witness by our behaviour at home, on the roads, in the workplace and so on. What a responsibility! What a privilege! Surely each one of us can accept the challenge to be more than babes in Christ? Then we are no longer drinking "milk" but eating "solid food." Then we are strong. But let us never forget, we are strong - **in the Lord.**

'For everyone who partakes only of milk is unskilled in the word of righteousness, for he is a babe.

But solid food belongs to those who are of full age, that is, those who have by reason of use of their senses exercised to discern both good and evil.'

Hebrews 5 verses 13 & 14

Has the Lord encouraged you as you have read this? Are you ready to take up this deeper challenge? Are you raring to go? Before leaping in, may I offer some words of caution? I speak from not a few years of experience or hearing few testimonies. We have an enemy who would spoil our resolve any way he can. His ways are well practiced and have been polished since the days of Adam and Eve. Yes, he will tempt us even through those who are closest to us.   He will try to tempt us in many other ways - especially where we are weak. Be warned, he will tempt us especially now we are working for the Lord. Let's face it; we were not much threat to him as babes. Now he will try to make us feel inadequate or, at the other end of the spectrum, proud. He will literally be hell bent on stopping us or spoiling our witness.   The media are quick to publicise any bad news, and jump at the chance to shame Christians!   Yes, the devil was defeated at Calvary, but he is still alive and kicking on planet Earth! We **must** ask for God's protection for self, family and even pets; for homes, travelling and relationships; for the church and each other. We **must** put on the whole armour of God. Have you noticed it is frontal armour?   We're to advance not retreat!   Yet ask God to reveal vulnerable areas; pray; and be prepared to fast if directed.

If for no other reasons than for those mentioned in the previous paragraph, this work should not to be tackled alone! I commend you to join a group of like-minded Christians. How do you find such a group of believers? You will know them by their fruits.

> *"Therefore by their fruits you will know them.*
>
> *Not everyone who says to Me, 'Lord, Lord,' shall enter the kingdom of heaven, but he who does the will of My Father in heaven.*
>
> *Many will say to me on that day, 'Lord, Lord, have we not prophesied in Your name, cast out demons in Your name, and done many wonders in Your name?'*
>
> *And then I will declare to them, 'I never knew you; depart from Me, you who practice lawlessness!"*
>
> Matthew 7 verses 20 to 23

By that scripture we can see that we should not believe in others **only** by the works they do! No, not even if a prophecy has been uttered, or some wonderful deed performed!

> *"For false Christ's and false prophets will rise and show great signs and wonders to deceive, if possible, even the elect."*
>
> Matthew 24 verse 24

I am reminded of 1 Corinthians chapter 13.

> *'Though I speak with the tongues of men and of angels, but have not love, I have become sounding brass or a clanging cymbal.*

And though I have the gift of prophesy, and understand all mysteries and all knowledge, and though I have all faith, so that I could remove mountains, but have not love, I am nothing.

And though I bestow all my goods to feed the poor, and though I give my body to be burned, but have not love, it profits me nothing.

Love suffers long and is kind; love does not envy;

Love does not parade itself, is not puffed up;

Does not behave itself rudely, does not seek its own, is not provoked, thinks no evil;

Does not rejoice in iniquity, but rejoices in truth;

Bears all things, believes all things hopes all things, endures all things.

Love never fails.   But whether there are prophecies, they will fail; whether there are tongues, they will cease; whether there is knowledge, it will vanish away.

For we know in part and we prophesy in part.

But when that which is perfect has come, then that which is in part will be done away.

When I was a child, I spoke as a child, I understood as a child, I thought as a child; but when I became a man, I put away childish things.

For now we see in a mirror, dimly, but then face to face. Now I know in part, but then I shall know just as I am also known.

And now abide faith, hope, love, these three; but the greatest of these is love.'

What a tremendous chapter St. Paul wrote! Surely **love** is the yardstick by which our hearts are tested?

Now our intention has been justified, let us consider the practicality of being in a group concerned with going further. If you can't find one, may I suggest that **you** form one? Why not? Don't listen to the enemy. Would not the result of forming such a group bless others and cause the Lord to smile upon you? Would you not desire this? This is dynamic! This is exciting! Pray for God's guidance and blessing, and then invite committed Christians and form one. May I offer the suggestion that it could be called an Omega Group? Is this presumption? I think not. I think that God has laid on my heart the present-day need for mature Christians to wait upon the Lord, to be filled with the Holy Spirit, then to step forward manifesting the gifts and fruit of the Holy Spirit. Is this not what a follower of Jesus should do? Is this not discipleship?

The challenge, then, is to have groups, (may we call them Omega Groups?), who feel there is more, yet in humility will not proceed in their own strength, but feel the need to 'wait on the Lord'.

> *'Even the youths shall faint and be weary, and the young men shall utterly fall.*
>
> *But those who wait on the Lord shall renew their strength;*
>
> *They shall mount up with wings like eagles,*
>
> *They shall run and not be weary,*
>
> *They shall walk and not faint.'*
>
> Isaiah 40 verse 30 & 31

Yes, waiting on the Lord, and having an open Bible[18], are both absolutely necessary. We should

---

[18] Just as by reading a map the mountain walker is warned not to step over a cliff, so the Bible is necessary to the searching believer to guide against dangers. It would be folly indeed, if the group went in the wrong direction because what was thought as 'guidance' was not checked out with God's Word. Is that not how some cults go astray?

be willing to listen to, learn from, and obey, God. Every group will be different. There will be much variety. That is why I have not here presumed to, *and feel no one should*, propose a syllabus for Omega Groups! Surely the Holy Spirit will do that - as each individual group meets. And no one will do a better job than He! God alone knows the different needs for each, and from each! God alone can give His grace and send the Holy Spirit to inspire what is needed, where it is needed, and when it is needed! We just have to listen, be open to Him, and be available - the highest service we can offer God!

Notwithstanding what I said in the last paragraph, perhaps if you gather to form an Omega Group, before praying, you can consider this:

'Even so you, since you are zealous for spiritual gifts,

Let it be for the edification of the church that you seek to excel.'

1 Corinthians 14 verse 12

CONCLUSION

It has not been my intention in this book to write great volumes. Suffice it to challenge our commitment and whet our appetite to grow up as Christians - Christians who will not have just been saved but who:

Wait on the Lord,
Exhibit the fruit of the Holy Spirit,
Exercise spiritual gifts,
Draw many into the kingdom of God,
Encourage and comfort fellow Christians,
Know what to do about sickness and demons.

But be on your guard.

'Be sober, be vigilant; because your adversary the devil walks about like a roaring lion, seeking whom he may devour.

Resist him, steadfast in the faith, knowing that the same sufferings are experienced by your brotherhood in the world.'
Peter 5 verses 8 & 9

Because even the most devout will witness temptations, including subtle complaints and disputes intended to make us 'break ranks'.

> *'Do all things without complaining and disputing,*
>
> *That you may become blameless and harmless, children of God without fault in the midst of a crooked and perverse generation, among whom you shine as lights in the world.'*
>
> Philippians 2 verses 14 & 15

May your light so shine as an Omega Christian!

May our heavenly Father enable you,
Through Jesus Christ, our Lord and Saviour,
And by the power of the Holy Spirit.
Amen.

PART TWO

From
DECEPTION
To
DELIVERANCE

All Scripture References are from
The New King James Version
THE HOLY BIBLE

INTRODUCTION (PART TWO)

I mentioned in part one that 'I believe if we allow the enemy to get established in our thoughts, we can be diverted from God's perfect plan for our lives'. I felt that this subject needed to be researched further, and should be separate from the topics looked at in that part of this book.

We will see in this part that there are many ways in which we can be deceived. Over and above these there are many more. Sicknesses, diseases, temptations; our archenemy has a whole repertoire, and is well practiced at using them. By deception we can be hoodwinked into doing things we would not normally do, and lose the peace of God and effectiveness in our lives.

We need to know what can be done about it. I think there are five main areas:

Firstly, we should 'keep our slate clean' to be effective. If we have an area needing forgiveness, we must come before the Lord to that end.

> *'Confess your trespasses to one another, and pray for one another, that you may be healed. The effective, fervent prayer of a righteous man avails much.'*
> James 5 verse 16

Secondly, we should ask for discernment, so that spiritually we are aware of what to come against, or of what is coming against us.

Thirdly, we may be led to fast as well as pray. (There are some who regularly fast for that area to be covered. They do this so that they are not taken by surprise when needed for a special case.)

> *"However, this kind does not go out except by prayer and fasting."*
> Matthew 17 verse 21

Fourthly, we need to put on the armour of God (if we have not done so already).

Lastly (Actually really a continuation of the fourth area above), we need to take the sword of the Spirit, which is the Word of God, and come against the sickness or evil spirit with scripture and, "In the name of Jesus Christ."

Chapter Nine

HOW IT ALL STARTED

In Genesis we read of the creation, and are told that God was well pleased with all He had made. The last verse in the first chapter of the Bible:

'Then God saw everything that He had made, and indeed it was very good. So the evening and the morning were the sixth day.'
Genesis 1 verse 31

Twenty-six verses later, we read of the serpent's cunning. The rot sets in! In just five further verses scripture states the first deception to mankind has been formulated, the hook has been baited, and the bait taken. God finds out (I think He already knew), and the blame game starts:

'Then the man said, "The woman whom You

gave to be with me, she gave me of the tree, and I ate."

And the Lord God said to the woman, "What is this you have done?" The woman said, "The serpent deceived me, and I ate."'
Genesis 3 verses 12 &13

So, we have now - deception, a lie, disobedience and trying to blame another for one's own shortcomings. All this in the first three chapters of the Bible! And it started so well! But it was not to end there. Before chapter six is half way through, it was not just Adam, Eve and Cain that were disappointing God!

'Then the Lord saw that the wickedness of man was great in the earth, and that every intent of the thoughts of his heart was only evil continually.

And the Lord was sorry that He had made man on the earth, and He was grieved in His heart.'
Genesis 6 verses 5 & 6

God was so grieved that He pronounced His destruction plan, and introduced His regeneration plan using Noah (who had found grace in the Lord's eyes, verse 8). We can be grateful that

Noah did not choose to disobey what he was then told to do. We can be grateful that Noah ignored the mocking and discouragement of many, and stuck to his resolve to continue with what God had asked him.

Chapter Ten

SALVATION

"So it will be at the end of the age. The angels will come forth, separate the wicked from among the just, and cast them into the furnace of fire. There will be wailing and gnashing of teeth."
Matthew 13 verses 49 & 50

Jesus is saying that there is punishment for those who have done wrong. That means all of us.

'For all have sinned and fall short of the glory of God.'
Romans 3 verse 23

Here is a story of two young men who were very good friends at school, but over the passage of time they had lost contact with each other.

Many years later circumstances brought them

together. One had become a judge, but the other a thief. They now came face to face with each other. The thief thought that because the judge and he had shared such good times, he would be merciful for 'old time's sake'.

"Fined £500, or six months in prison." He was not a little surprised; not what he expected at all. He couldn't pay, so down to the cells he was taken.

After all the court appearances, there were footsteps on the stairway leading down to the cells. It was the judge. "Fine friend you turned out to be, at least you could have given me a suspended sentence." said the thief. "Justice is necessary," said the judge, "but what I will do as your friend, on this occasion only, is pay the fine for you." I think you will agree the judge was both wise and generous.

So, in the eyes of God, we are like that thief. In fact we have 'stolen' our lives from Him to do our own thing. He had good lives planned for us, but we took it upon ourselves to sin one way or another. Now Jesus is like that judge. He has paid the price for our sin.

'But God demonstrates His love towards us, in that while we were still sinners, Christ died for

us.'
                               Romans 5 verse 8

'For the wages of sin is death, but the gift of God is eternal life in Christ Jesus our Lord.'
                              Romans 6 verse 23

'For Christ is the end of the law for righteousness to everyone who believes.'
                              Romans 10 verse 4

'That if you confess with your mouth the Lord Jesus and believe in your heart that God has raised Him from the dead, you will be saved.'
                              Romans 10 verse 9

To summarise - hell was our destiny because of sin, but we can be saved from it through Jesus Christ. If we know some who are not assured of personal salvation, we should get them to pray this or similar:

"Heavenly Father, I have sinned and have offended you.

Please forgive me.

I now know Jesus; You died to take my punishment upon Yourself.

Jesus, thank you so much.

I believe Jesus; You rose from the dead and want to be my Saviour.

Jesus, not only save me, but also come into my life to help me live a life pleasing to our heavenly Father.

I've made a mess of my life by my efforts; Show me the way; be my Lord.
Amen."

If they have genuinely prayed that, assure them that they are now Christians; that they have made the most important decision of their lives; that there are even angels in heaven rejoicing over that decision.   Welcome them as your brothers or sisters in Christ.

The above may be obvious but, because there is so much deception, they may be tempted to doubt their salvation.   That is why I suggest the above, and this word of encouragement

> *'No weapon formed against you shall prosper,*
> *And every tongue which rises against you in judgment*
> *You shall condemn.*
> *This is the heritage of the servants of the Lord,*
> *And their righteousness is from Me,*
> *Says the Lord.'*
>   Isaiah 54 verse 17

So we see that when our thoughts are negative or ungodly, we can condemn them. This is our heritage as Christians.

Chapter eleven

SOME THOUGHTS

I feel we should take a quick look at the subject of thoughts, as this is the channel through which we can be deceived. Thoughts – that process that goes on in our heads; that activity which precedes our actions. Well, mostly, anyway. We need not look into spontaneous action as a result of pain or danger – you don't have to think about moving your hand forward to break a fall, for example. Here we need to think about, well… thinking. Thoughts can help us, or hinder us; they can keep us out of trouble, or get us into trouble; and they can be changed by outside circumstances, or guided in another way.

We might consider our minds like radio sets - with different channels we can tune into. One has our own interests at heart (Like, "I'm hungry, I'd better get a meal"); another has good intentions for others at heart (like, "I think that old lady needs to cross the road, I'll offer to help her");

yet another has bad intentions (like, "You like that handbag, Why don't you steal it?"[19]). We can choose, if you like, which channel to switch on; and we can choose when to switch off; we can choose to change channels. After all God has given mankind free will. Yes? Well, if only it were that simple! The channel we want may get interference. To get rid of that interference, we may well need help with our tuning. Spiritual interference may come in the form of temptation or deception. For spiritual interference we need spiritual help. We need grace to resist temptation; we need discernment to know deception for what it is; and we need wisdom to know how to deal with spiritual problems.

This is where this study comes in. Our minds are under attack. After all, is it not the mind that is the battlefield? We can be tempted or deceived into thinking that something is OK when, in fact, it is far from that. We need deliverance from the temptation before we fall for it. If we miss that, then we will need deliverance and forgiveness afterwards.

I hope, in the following chapters, with God's grace and help, you will find encouragement so that you are able to resist and combat temptation and deception.

---

[19] Wondering about that one? Now, where do you think that thought came from?

Chapter Twelve

MORE ABOUT THOUGHTS

Regarding our thoughts:

*'You will keep him in perfect peace,
Whose mind is stayed on You.
Because he trusts in You.'*
                    Isaiah 26 verse 3

Later in Isaiah we read about God's thoughts:

*"For My thoughts are not your thoughts,
Nor are your ways My ways," says the Lord.
"For as the heavens are higher than the earth,
So are My ways higher than your ways,
And My thoughts than your thoughts."*
                    Isaiah 55 verses 8 & 9

We know the natural man's thoughts fall far short of what God would have us receive.

> *'But the natural man does not receive the things of the Spirit of God, for they are foolishness to him; nor can he know them because they are spiritually discerned.'*
> 1 Corinthians 2 verse 14

But we are not 'natural' when we have asked Jesus Christ to live in us and are filled with the Holy Spirit. Indeed if we continue to the following verses, we read:

> *'But he who is spiritual judges all things, yet he himself is rightly judged by no one. For 'who has known the mind of the Lord that he may instruct Him?'* [20]

> *'But we have the mind of Christ.'*
> 1 Corinthian 2 verses 15 & 16

Now, to whom is the 'we' being referred to? If we look deeper we read:

> *'For the message of the cross is foolishness to those who are perishing, but to us who are being saved it is the power of God.'*
> 1 Corinthians 1 verse 18

---

[20] From Isaiah chapter 40 verse 13.

> *'And I, brethren, could not speak to you as to spiritual people but as to carnal, as to babes in Christ.'*
>
> 1 Corinthians 3 verse 1

From these, it would seem that 'we' referred to those who are saved and spiritual. Paul here also seems to put great emphasis on the need to have overcome worldly (carnal) weaknesses. I offer this remark in light of what follows from verse one:

> *'I fed you with milk and not solid food; for until now you were not able to receive it, and even now you are still not able; for you are still carnal.*
>
> *For where there are envy, strife, and divisions among you, are you not carnal and behaving like mere men?'*
>
> 1 Corinthians 3 verses 2 to 3

So is Paul saying that, to have the mind of Christ we would not be behaving like mere men?

To summarise the above scriptures from Corinthians, can we conclude that to have the mind of Christ we have to be saved, filled with the Holy Spirit, and no longer have envy, strife and divisions among us? Would we then fulfil the conditions necessary to have the mind of Christ?

I have read a simpler suggestion, 'that we have the mind of Christ because the Holy Spirit dwells in us, whence we would be filled to overflowing and have access to the all-knowing mind of Christ'. We may consider this. But before this goes to our head (no pun intended), let us remember whom we are. But for the grace of God, would not mankind have been wiped out long ago? If we dared to think that we could also be all knowing (after all *if* we were of the same mind…) we may think of ourselves equal to God? That would be heresy! The devil went down that route – and look where it got him!

There is another question about the words 'all knowing', and that hinges on not even Jesus being all knowing. Does not scripture state:

> *"But of that day and hour no one knows, not even the angels in heaven, nor the Son, but only the Father."*
>
> Mark 13 verse 32

But why are we considering 'having the mind of Christ'? Well, if we had His mind, there would be no question about whether there was deception, or, whether deliverance was necessary. He knew exactly what was afoot - and what to do about it. If we had the same mind, well, would we not also know how to correctly deal with all situations?

Nevertheless I think we have to be very careful about what we wish to claim. We can become too clever for our own good. Pride is definitely out for the true Christian. I believe it is the door through which false Christ's and false prophets come to us.

So, considering all the above, what really is Paul meaning by having the mind of Christ? Could it simply mean thinking to do the will of God? Period! Now, there's a thought (and the pun is intended)!

Carnal Christians would then be those who would want to do this, but who still have areas where Jesus is not Lord. I am sure you have heard the saying of considering ourselves as a house. As Christians we have invited Jesus through our front door, but is He allowed in every room? That includes even the basement, attic and larder. Is He really Lord of all?

In conclusion about these thoughts, rather than thinking we are 'high and mighty', and able to do all He does, should we not show our gratitude for all God has done for us? We should remember all that Jesus suffered for us. Could we, indeed, do all He did? Remember although

Jesus was tempted much, He would think no evil at all. And did He not make the heavens? Now where do we stand? We should be humble before our God, and offer ourselves to be used in whatever way He knows best!

> *'Therefore do not be unwise, but understand what the will of the Lord is.'*
>
> Ephesians 5 verse 17

> *'Finally, brethren,*
> *Whatever things are true,*
> *Whatever things are noble,*
> *Whatever things are just,*
> *Whatever things are pure,*
> *Whatever things are lovely,*
> *Whatever things are of good report,*
> *If there is any virtue and if there is anything praiseworthy,*
> *Meditate on these things.'*
>
> Philippians 4 verse 8

Chapter Thirteen

THE SPOKEN WORD

The spoken word. And what is necessary for that? The tongue, of course! And what controls the tongue? Thoughts! Here we go again; just when we *thought* that subject was finished. Thoughts have a lot to answer for. I say again, thoughts have a lot to answer for - or should it be the tongue? No, the tongue has not been used to convey these words. There again, is not the tongue a little part of the body controlled by thoughts? Before I get you mixed up, let's turn to what some scriptures record on the subject. May the Lord reveal to each of us what he wants us to know.

> *'For we all stumble in many things. If anyone does not stumble in word, he is a perfect man, able to bridle the whole body.*
>
> *Indeed, we put bits in horses' mouths that they may obey us, and we turn the whole body.*

Look also at ships: although they are so large and are driven by fierce winds, they are turned by a very small rudder wherever the pilot desires.

Even so the tongue is a little member and boasts great things.   See how great a forest a little fire kindles!

And the tongue is a fire, a world of iniquity. The tongue is so set among our members that it defiles the whole body, and sets on fire the course of nature; and is set on fire by hell.

For every kind of beast and bird, of reptile and creature of the sea, is tamed and has been tamed by mankind.

But no man can tame the tongue.   It is an unruly evil, full of deadly poison.

With it we bless our God and Father, and with it we curse men, who have been made in the similitude of God.

Out of the same mouth proceed blessing and cursing.   My brethren, these things ought not be so.'

James 3 verses 2 to 10

Strong words, but what about these words

from James even earlier:

> *'If anyone among you thinks he is religious, and does not bridle his tongue but deceives his own heart, this one's religion is useless.'*
> James 2 verse 26

All this does not say very much for the tongue. But that is not to say all our speech is bad. Even above it states that we speak blessings as well as other things. And it does say, 'No **man** can tame…'[21] Indeed being filled with the Holy Spirit changes all that.

To be good witnesses, we just have to watch what we say. If we find this a problem area, we shall just have to ask the Lord to 'put a watch over our mouths'.

'And does not bridle his tongue'[22], implies that we are able to control what we say.

'Set on fire by hell'[23], and, 'an unruly evil', are both stating an involvement by the enemy. For this to happen, is it not implied that the enemy has got into our thoughts? Perhaps what we have said was defensive, and a result of accusation whence our pride had been hurt. Is

---

[21] 'No man can tame' is mentioned in the James 3 quote
[22] 'Bridle his tongue is' mentioned in the James 2 quote.
[23] 'Set on fire by hell' and 'an unruly evil' are both mentioned also in the James 3 quote.

not pride a problem area in itself? Or we may have been rebellious. The Bible has something to say against rebellion:

> *'For rebellion is as the sin of witchcraft,*
> *And stubbornness is as iniquity and idolatry.*
> *Because you have rejected the word of the Lord,*
> *He also has rejected you from being king.'*
> 1 Samuel 15 verse 23

There are many ways we can be deceived into thinking that we have a right to speak, or answer back, the way we do.

It may be wise to remember:

> *'A soft answer turns away wrath, but a harsh word stirs up anger.*
>
> *The tongue of the wise uses knowledge rightly,*
>
> *But the mouth of fools pours forth foolishness.'*
> Proverbs 15 verses 1 & 2

There is even a worldly saying that says something about 'counting up to ten'. Perhaps some of us need to count more than that! And how about praying whilst counting. Can't do both? Well then, just pray.

Chapter Fourteen

THE WRITTEN WORD

Of course, without the written word, you would not be reading this. And within this study I have been quoting other written words – those of scripture. Scriptures – those words recorded in The Holy Bible - God's Word.

One verse from the Old Testament:

'Your word is a lamp to my feet and a light to my path'

Psalm 119 verse 105

And one from The New Testament:

'The word of God is living and powerful, and sharper than any two-edged sword, piercing even to the division of soul and spirit, and of joints and marrow, and is a discerner of the thoughts and intents of the heart.'

Hebrews 4 verse 12

I like one person's remark about The Holy Bible. Harold Hill, who wrote amongst others, 'How To Live Like A King's Kid', calls the Bible, 'The Manufacturer's Handbook'. By the way he also calls our brains, 'educated idiot boxes'. Maybe there is food for thought about that remark.

This is what 'The Manufacturer's Handbook' also says about itself:

> *'All scripture is given by inspiration of God, and is profitable for doctrine, for reproof, for correction, for instruction in righteousness,*
>
> *That the man of God may be complete, thoroughly equipped for every good work.'*
> 2 Timothy 3 verses 16 & 17

Studying the written word is a vast subject. I am going to include just one more verse, before closing this chapter on its importance:

> *'And having been perfected, He became the author of eternal salvation to all who obey Him.'*
> Hebrews 5 verse 9

Chapter Fifteen

WHAT HINDERS

Perhaps the main reason why we may be reluctant to move forward to do things the Lord asks of us is lack of faith. Some have none at all:

'Finally, brethren, pray for us, that the word of the Lord run swiftly and be glorified, just as it is with you,

And that we may be delivered from unreasonable and wicked men; for not all have faith.'
2 Thessalonians 3 verses 1&2

'For not all have faith'. So that does not refer to 'the faithful' i.e. to us. Nevertheless some of us have just a little faith. (So we know what to do):

'So then faith comes by hearing, and hearing by the word of God.'
Romans 10 verse 17

Others are hindered by those around them. Even Jesus found this:

> *'Then He came to the house of the ruler of the Synagogue, and saw a tumult and those who wept and wailed loudly.*
>
> *When He came in, He said to them, "Why make this commotion and weep? The child is not dead, but sleeping."*
>
> *And they ridiculed Him. But when He had put them all outside, He took the father and the mother of the child, and those who were with Him, and entered where the child was lying.*
>
> *He then took the child by the hand, and said to her, "Talitha, cumi," which is translated, "Little girl, I say to you, arise."*
>
> *Immediately the girl arose, and walked, for she was twelve years of age. And they were overcome with amazement.'*
>
> Mark 5 verses 38 to 42

Just a few verses later:

> *'Is this not the carpenter, the Son of Mary, and brother of James, Joses, Judas and Simon? And are not His sisters here with us? So they*

> *were offended at Him.*
>
> *But Jesus said to them, "A prophet is not without honor except in his own country, among his own relatives, and in his own house."*
>
> *Now He could do no mighty work except that He laid His hands on a few sick people and healed them.'*
>
> Mark 6 verses 3 to 5

But have you noticed that, even then, a few were healed? Before being of use, the lesson here is, of course, to draw away from those who discourage - or worse.

I have mentioned above a possible main reason for being unable to move forward, being 'lack of faith'. But what if that is not the main reason? What if the main reason was because of what is within us? These scriptures show what I mean:

> *'These six things the Lord hates,*
> *Yes, seven are an abomination to Him:*
> *A proud look,*
> *A lying tongue,*
> *Hands that shed innocent blood,*
> *A heart that devises wicked plans,*
> *Feet that are swift in running to evil,*

A false witness who speaks lies,
And one who sows discord among brethren.'
Proverbs 6 verses 16 to 19

'Therefore put to death your members which are on earth: fornication, uncleanness, passion, evil desire, and covetousness, which is idolatry.'
Colossians 3 verse 5

'Therefore we also, since we are surrounded by so great a cloud of witnesses, let us lay aside every weight, and the sin which so easily ensnares us, and let us run the race that is set before us.'
Hebrews 12 verse 1

Then there is Galatians chapter 5, and Ephesians chapter 5, and more, if required.

The main thing is that we should keep away from those who would hold us back, and walk in goodness and righteousness to be effective.

Chapter Sixteen

DELIVERANCE

Before tackling this subject, I feel we must ask for protection.

We ask the Lord for protection, and put on our spiritual armour:

> *'Put on the whole armour of God, that you may be able to stand against the wiles of the devil.*
>
> *For we do not wrestle against flesh and blood, but against principalities, against powers, against the rulers of the darkness of this age, against spiritual hosts of wickedness in the heavenly places.*
>
> *Therefore take up the whole armour of God, that you may be able to withstand in the evil day, and having done all to stand.*
>
> *Stand therefore, having girded your waist with truth, having put on the breastplate of*

righteousness,

And having shod your feet with the preparation of the gospel of peace;

Above all, taking the shield of faith with which you will be able to quench all the fiery darts of the wicked one.

And take the helmet of salvation, and the sword of the Spirit, which is the word of God.'
Ephesians 6 verses 11 to 17

In an earlier chapter, I mentioned the many, the some and the few, who were involved with preaching, healing and casting out demons. I implied the last two appeared to need more attention. This chapter is concerned with these two needs.

It amazes me to hear so many people blame God when something has gone wrong. Disasters can be called, 'an act of God' by insurance companies. Death of a loved one and sicknesses can be regarded as being allowed by an uncaring God. Some even lose faith at this point. Has it not occurred to so many of us that it is not God but the devil that is responsible for so much misery? Even if the heartache has come through a natural cause, the enemy tries to sow seeds of discontent against God, and disbelief in God.

Some then doubt His love. Yes, sickness and demonic activity are much used by the devil to deceive and cause misery.

God does not want us to think He is uncaring, nor does He want us to suffer with sickness and demonic possession. Why else would Jesus have sent out His disciples to heal the sick and cast out demons? Why else would there be so many examples of deliverance by Him, His disciples and believers, in the gospels, Acts of the Apostles and many letters?

I think there is a huge amount of sickness in the world today. I also think that deception is rife. I've even heard it said that demonic activity finished a long time ago! Let's just think of spiritual darkness, deception, doubt, despair, depression, disobedience, desperation, delusion, delinquency, destruction and denial. Note all the d's? Now, if we put a d in front of evil, we get devil. Well, yes, there are some good d's. But can you see the enemy can be to the forefront of the examples I have given? I'm sure you could find a few more. Furthermore, I think that both sickness and the influence of evil spirits could be dramatically reduced if more Christians took the Lord's commission seriously. Would not the waiting lists at hospitals be reduced? Would not the bad behaviour of so many be overturned?

We should not just sit back, think we are saved, and do nothing spiritually until life for us here is finished. We have a job to do!

If that does not challenge you, please think what Jesus did for us:

> *'Surely He has borne our griefs and carried our sorrows:*
> *Yet we esteemed Him stricken, smitten by God, and afflicted.*
> *But He was wounded for our transgressions,*
> *He was bruised for our iniquities;*
> *The chastisement for our peace was upon Him,*
> *And by His stripes we are healed.'*
> Isaiah 53 verses 4 & 5

Now, surely we can help a few sick people by saying -

"In the name of Jesus Christ, be healed."

And what about relieving someone from a demon's influence (e.g. alcoholism, nicotine, or other drugs)? "In the name of Jesus Christ."

Well, if some people, who were not known to Jesus, could cast out, why shouldn't we?

How do I justify that remark?

> *"Many will say to me on that day, 'Lord, Lord, have we not prophesied in Your name, <u>cast out</u>*

demons in Your name, and done many wonders in Your name?'

And then I will declare to them, 'I never knew you; depart from Me, you who practice lawlessness!"
Matthew 7 verses 20 to 23

"So in the name of Jesus and by His stripes we are healed, we understand." You say. "But casting out demons; and one of them possibly being called 'Nicotine', you can't be serious?" You may not agree, but let us research further.

Over four hundred years ago, according to the Internet, king James 1 regarded smoking as 'a sin against God' and 'the fumes likened to those rising from hell'. Since then trillions of pounds sterling have been spent on tobacco, billions on treatments and lost work through sickness, and millions have suffered lengthy ill health, many of who died prematurely in a very distressing way. We are now told that many more will suffer as a result of even being near a smoker, or to enter a room in which smoking has taken place. Even unborn babies can suffer. Let's face it, if the devil had nothing to do with all that, he sure would have wished to employ the one who had. By the way, have I mentioned 'deception' earlier?

Now what are you going to say to me? Maybe,

"You don't know what you're talking about." "It's all right for you; speaking like that you have never smoked." I'm sure many other remarks could be made. Let's cut it short. I too have smoked. I remember being the only smoker in six on top of a mountain when supplies including 200 cigarettes were dropped by parachute. Wow, 200, just for me - I remember in the two days I smoked a total of - half a cigarette (the air was thin); just wasn't the same! Sometimes I smoked Capstan Full Strength; even tried pipe tobacco - yes, rolled up into a cigarette! Mind you, wasn't too keen! In all I smoked for about twenty years, finishing for good more than thirty years ago, thanks be to God. So I **do** know what it is like. And I **do** know Nicotine does not have to keep you bound! Indeed, we can bind it and cast it out, in the name of Jesus Christ!

I have mentioned 'deception' before, haven't I? Well there are many ways we can be deceived. We can be deceived into believing there is no such thing as demons, or if there are, only specialist Christians can do something about them[24]. Mankind can be deceived into 'playing' with the occult – anything from horoscopes to Satanism. We can be deceived into thinking we are standing

---

[24] Not so! See end of Mark's gospel if you doubt. See also 'WARNING' at end of this chapter on deliverance.

up for our rights, when in fact a controlling spirit is attempting to establish itself. That last is an example of a whole host of demons that want to invade us humans if they get a chance to enter by a sin like pride. I am not going to glorify the devil by giving a long list of different names for these demons or evil spirits. Nor am I going to cause you concern by so doing. You may feel overwhelmed even now, so I will immediately give you these promises:

> *'But He gives more grace. Therefore He says:*
>
> *"God resists the proud,*
> *But gives grace to the humble."*
>
> *'Therefore submit to God. <u>Resist the devil and he will flee from you</u>.'*
> James 4 verses 6 & 7

And remember:

> *'There is no fear in love; but perfect love casts out fear, because fear involves torment. But he who fears has not been made perfect in love.*
>
> *We love Him because He first loved us.'*
> 1 John 4 verses 18 & 19

Later in the same epistle:

> *'By this we know that we love the children*

of God, when we love God and keep His commandments.

For this is the love of God, that we keep His commandments, and His commandments are not burdensome.

For whatever is born of God overcomes the world. And this is the victory that has overcome the world - our faith.

Who is he who overcomes the world, but he who believes that Jesus is the Son of God?'
1 John 5 verses 2 to 5

And, the final quote:

'You will keep him in perfect peace,
Whose mind is stayed on You,
Because he trusts in You.'
Isaiah 26 verse 3

Finally, a WARNING:

In deliverance ministry it is unwise to be alone. Remember even Jesus sent the disciples out in two's! You would be well advised to have others (even if miles away) praying prayers of protection. I am sure that Jesus covered His disciples in prayer as they went on their way.

PART THREE

TWO PROPHESIES
And
EPILOGUE

PROPHESY FOR TODAY - NO 1

"You are My friends. Yes, I call you friends, says the Lord. Some of you have known me a long time, some a little, some just. But you are My friends, Our Heavenly Father and I love you, and We desire the Holy Spirit to fill you.

You have looked towards others for learning and example. That is good, but I also want you to look to Me. Copying others' good works is not enough. Each one of you is unique, and the purpose I have for each one of you is unique. Remember how different each one of you is; not even 'identical' twins have the same fingerprints.

But many unique children of God are heading for destruction. The prince of darkness wants to destroy as many as he can. The days are now short. The end time is near. Action is needed now as never before. Be My soldiers. Put on the Heavenly Armour and fight. Be not faint hearted, says the Lord. Fight!

But this fight is not what many of you think. Yes, there will be persecution, and there will be more martyrs in the world. But the fight I am talking about is for all who call themselves "Christian". This fight you will find impossible to win without Me. Take courage and be filled by The Holy Spirit. Let the fruit of The Spirit spring forth in your lives. Show those around you what is Christ-in-ity.

Be not deterred by ridicule, laughter or other offences. Remember by being strong in this - your example will help to save lost souls."

PROPHESY FOR TODAY - NO 2

"There is a great darkness descending over the world. Evil is becoming much more evident and is spreading across the nations.

There will be many who will look towards My people for help in this age.

You must be ready, for you are the lights in the darkness. You must discard the old oil that is in your lamps because it has become contaminated with water, and renew with pure, fresh oil. You must renew the wicks so that they last the duration. Then you must go out holding your lamps high, so that shadows will not come between you and those who wish to see the way.

Do not concern yourselves with those who turn their backs on you to face the darkness, for the time is short, and others need to see the light.

Pray that your lamps do not dim, or become extinguished, for I shall be holding you to account."

EPILOGUE

Once I was asked to give a talk at a F.G.B.M.F.I.[25] meeting. The flyer stated that I had been 'kept' for such an occasion.

Could it be that I have been 'kept' so that I could write, 'The Christian Challenge From Alpha to Omega'?

If you consider this possible, would you also consider what importance for you the Lord places on what has been written?

Normally when one comes to the end of a book it is marked:

The End

However, in this case, I really do feel it should be marked:

THE BEGINNING

Of your new life as an Omega Christian.

(If you were not one already).

Now it's over to you.

God bless.

[25] The Full Gospel Business Men's Fellowship International is a non-denominational organisation, which invites non-believers to a meal. A speaker then gives testimony after which there can be a time for ministry and prayer. Many have come to know the Lord through the testimonies, and there have been many healings witnessed. Reading, 'The Happiest People on Earth', you can discover how this started.

Milton Keynes UK
Ingram Content Group UK Ltd.
UKHW041906040924
447889UK00001B/7

9 781434 325297